In Our Country

Susan Canizares • Daniel Moreton

Scholastic Inc.
New York • Toronto • London • Auckland • Sydney

Acknowledgments

Literary Specialist: Linda Cornwell
Social Studies Consultant: Barbara Schubert, Ph.D.

Design: Silver Editions
Photo Research: Silver Editions
Endnotes: Jacqueline Smith
Endnote Illustrations: Anthony Carnabucia

Photographs: Cover: (tl) Stephen G. Maka/The Picture Cube, Inc.; (tm) Collins/Monkmeyer Press; (tr) Vanessa Vick/Photo Researchers, Inc.; (bl) Kenneth W. Fink/Photo Researchers, Inc.; (bm) Jim Erickson/The Stock Market; (br) Jim Foster/The Stock Market; p. 1: Charles Thatcher/Tony Stone Images; p.2: Vanessa Vick/Photo Researchers, Inc.; p. 3: Jim Erickson/The Stock Market; p. 4: Frank L. Simonetti/The Picture Cube, Inc.; p. 5: Stephen G. Maka/The Picture Cube, Inc.; pp. 6–7: John Buttenkant/Photo Researchers, Inc.; p. 8: Roy Corral/Tony Stone Images; p. 9: Kenneth W. Fink/Photo Researchers, Inc.; p. 10: Collins/Monkmeyer Press; p. 11: Jim Foster/The Stock Market; p. 12: Daemmrich/The Image Works.

No part of this publication may be reproduced in whole or in part, or stored in a retrieval system, or transmitted in any form or by any means, electronic, mechanical, photocopying, recording, or otherwise, without written permission of the publisher. For information regarding permission, write to Scholastic Inc., 555 Broadway, New York, NY 10012.

Library of Congress Cataloging-in-Publication Data
Canizares, Susan 1960-
In our country/Susan Canizares, Daniel Moreton.
p. cm. --(Social studies emergent readers)
Summary: Simple text and photographs explore the beauty and diversity of the United States, including its forests, deserts, beaches, marshes, and mountains.
ISBN 0-439-04562-2 (pbk.: alk. paper)
1. Physical geography--United States--Juvenile literature.
[1. Physical geography.] I. Moreton, Daniel. II. Title. III. Series.
GB121.C25

557.3--dc21

99-12153
CIP

Copyright © 1999 by Scholastic Inc.
Illustrations copyright © 1999 by Scholastic Inc.
All rights reserved. Published by Scholastic Inc.
Printed in the U.S.A.

This is a map of our country.

In some parts, there are forests.

In others, there are deserts.

In some parts, there are beaches.

In others, there are marshes.

In some parts, there are rivers.

In others, there are lakes.

In some parts, there are canyons.

In others, there are prairies.

In some parts,

there are snow-covered mountains.

Let's take care of our whole country.

In Our Country

The United States of America is the fourth-largest country in the world. Because the U.S. is such a big country, it has many different kinds of natural features.

Forests The forests of New England are famous for maple syrup and fall leaves. The tropical forests of Florida are home to more than 600 types of animals. The Northwest forests provide much of the lumber we use for construction. Perhaps the most unusual forests in the U.S. are the redwood forests of California. Redwoods are the tallest living things on earth.

Deserts Deserts are dry areas with very little rainfall and very extreme temperatures. There are many deserts in the Southwest region of the U.S. One of the hottest regions in the world is the desert in Death Valley, California. Temperatures can reach 125 degrees Fahrenheit!

Beaches There are beautiful beaches all along the coasts of the U.S., but the most famous beaches are in Hawaii, California, and Florida. The state of Hawaii, which is a group of islands formed by volcanic activity, has beaches with black sand formed from lava. California is full of white, sandy beaches. And the peninsula of Florida has seemingly endless beaches along the Gulf of Mexico and the Atlantic Ocean.

Marshes Marshes, where tall grasses grow, form in low areas next to rivers or coasts and are often rich in plant and animal life. Saltwater marshes, like the one on the island of Nantucket, are common along the East Coast of the United States. There are also many freshwater marshes along the Mississippi River. The Florida Everglades is a mixture of freshwater swamps and marshes where hundreds of plants and animals live.

Mountains The Appalachian Mountains in the eastern U.S. are the oldest mountain range in North America. Formed millions of years ago, they have been worn down by wind and water. In the West there are younger and higher mountain ranges: the Rockies, the Cascades, and the Sierra Nevadas. Mt. McKinley in the Alaska range is the highest point in North America.

Rivers Hundreds of rivers crisscross the entire country. Many rivers are controlled by dams, such as the great Hoover Dam on the Colorado River. Dams direct the water flow, adjust the water level, and prevent flooding. Still, some rivers are difficult to control. The Mississippi River, the longest river in the U.S., floods nearly every year, causing great damage to homes, businesses, and farms.

Lakes Lakes dot the entire U.S., and the largest are the Great Lakes, on the northern border with Canada. These huge freshwater lakes, connected by natural and artificial channels, were the route by which settlers traveled from the East to rich farmlands in the Midwest. Today these lakes are still very important for shipping, fishing, and recreation.

Canyons A canyon forms when, over millions of years, a river cuts into rock, carving a deep valley with steep walls. Most canyons in the U.S. can be found in the West. The Colorado Plateaus in the Southwest contain some of the most famous canyons in the world, such as Bryce Canyon in Utah and the huge Grand Canyon in Arizona, a natural wonder that 4 million tourists visit every year.

Prairies Prairies are grasslands with extremely fertile soil. The Great Plains, which stretch across the midwestern region of the United States, are covered with prairie grass. The prairies of the Midwest are sometimes called "the breadbasket of the world" because they produce grain for many parts of the world.

Our country is rich in natural wonders. But much damage has been done—forests have been cut down, swamps have been drained, rivers have been polluted, and soil has been overused and turned to dust. Luckily, starting in 1872 with the creation of Yellowstone National Park, the government has been putting aside land and protecting our natural treasures from development.